Energy Sector Standard of the People's Republic of China

NB/T 10242-2019

Specification for classification and coding of inventory of land requisition for hydropower projects

水电工程建设征地实物指标分类编码规范

(English Translation)

China Water & Power Press

中国水利水电出版社

Beijing 2024

All rights reserved. No part of this publication may be reproduced, stored in a retrieval system, or transmitted in any form or by any means—electronic, mechanical, photocopying, recording or otherwise, without prior written permission of the publisher.

图书在版编目（CIP）数据

水电工程建设征地实物指标分类编码规范：NB/T 10242-2019 = Specification for classification and coding of inventory of land requisition for hydropower projects (NB/T 10242-2019)：英文 / 国家能源局发布. -- 北京：中国水利水电出版社，2024. 3. -- ISBN 978-7-5226-2804-2

Ⅰ. TV752-65

中国国家版本馆CIP数据核字第2024B023K2号

Energy Sector Standard of the People's Republic of China

中华人民共和国能源行业标准

Specification for classification and coding of inventory of land requisition for hydropower projects

水电工程建设征地实物指标分类编码规范

NB/T 10242-2019

(English Translation)

Issued by National Energy Administration of the People's Republic of China

国家能源局　发布

Translation organized by China Renewable Energy Engineering Institute

水电水利规划设计总院　组织翻译

Published by China Water & Power Press

中国水利水电出版社　出版发行

　　Tel: (+ 86 10) 68545888　68545874

　　sales@mwr.gov.cn

　　Account name: China Water & Power Press

　　Address: No.1, Yuyuantan Nanlu, Haidian District, Beijing 100038, China

　　http://www.waterpub.com.cn

中国水利水电出版社微机排版中心　排版

北京中献拓方科技发展有限公司　印刷

210mm×297mm　16开本　2.25印张　91千字

2024 年 3 月第 1 版　2024 年 3 月第 1 次印刷

Price（定价）：￥370.00

About English Translation

This English version is one of China's energy sector standard series in English. Its translation was organized by China Renewable Energy Engineering Institute authorized by National Energy Administration of the People's Republic of China in compliance with relevant procedures and stipulations. This English version was issued by National Energy Administration of the People's Republic of China in Announcement [2023] No. 8 dated December 28, 2023.

This version was translated from the Chinese Standard NB/T 10242-2019, *Specification for classification and coding of inventory of land requisition for hydropower projects*, published by China Water & Power Press. The copyright is reserved by National Energy Administration of the People's Republic of China. In the event of any discrepancy in the implementation, the Chinese version shall prevail.

Many thanks go to the staff from the relevant standard development organizations and those who have provided generous assistance in the translation and review process.

For further improvement of the English version, any comments and suggestions are welcome and should be addressed to:

China Renewable Energy Engineering Institute
No. 2 Beixiaojie, Liupukang, Xicheng District, Beijing 100120, China
Website: www.creei.cn

Translating organizations:

China Renewable Energy Engineering Institute

China Three Gorges Corporation

Translating staff:

LIU Zuxiong	LIU Can	ZHANG Guoping	LONG Yang
LIAO Guihua	LIAO Leiqiong	SHANG Yanguang	WU Juan
CAO Lingyan	LIU Xi	BIN Lishu	WEN Liangyou

Review panel members:

WANG Kui	China Renewable Energy Engineering Institute
LIU Hao	POWERCHINA Zhongnan Engineering Corporation Limited
ZHANG Ming	Tsinghua University
XU Zeping	China Institute of Water Resources and Hydropower Research
LIU Xiaofen	POWERCHINA Zhongnan Engineering Corporation Limited
LI Zhongjie	POWERCHINA Northwest Engineering Corporation Limited
QI Wen	POWERCHINA Beijing Engineering Corporation Limited
GAO Yan	POWERCHINA Beijing Engineering Corporation Limited
AN Zaizhan	China Renewable Energy Engineering Institute

National Energy Administration of the People's Republic of China

翻译出版说明

本译本为国家能源局委托水电水利规划设计总院按照有关程序和规定，统一组织翻译的能源行业标准英文版系列译本之一。2023年12月28日，国家能源局以2023年第8号公告予以公布。

本译本是根据中国水利水电出版社出版的《水电工程建设征地实物指标分类编码规范》NB/T 10242—2019翻译的，著作权归国家能源局所有。在使用过程中，如出现异议，以中文版为准。

本译本在翻译和审核过程中，本标准编制单位及编制组有关成员给予了积极协助。

为不断提高本译本的质量，欢迎使用者提出意见和建议，并反馈给水电水利规划设计总院。

地址：北京市西城区六铺炕北小街2号
邮编：100120
网址：www.creei.cn

本译本翻译单位：水电水利规划设计总院
中国长江三峡集团有限公司

本译本翻译人员：刘祖雄　刘　灿　张国平　龙　旸
廖贵华　廖磊琼　商艳光　吴　娟
曹凌燕　刘　曦　宾莉姝　文良友

本译本审核人员：

王　奎　水电水利规划设计总院

刘　昊　中国电建集团中南勘测设计研究院有限公司

张　明　清华大学

徐泽平　中国水利水电科学研究院

刘小芬　中国电建集团中南勘测设计研究院有限公司

李仲杰　中国电建集团西北勘测设计研究院有限公司

齐　文　中国电建集团北京勘测设计研究院有限公司

高　燕　中国电建集团北京勘测设计研究院有限公司

安再展　水电水利规划设计总院

国家能源局

Contents

Foreword ··· VII
Introduction ··· IX
1　Scope ··· 1
2　Normative references ·· 1
3　General provisions ·· 1
4　Basic requirements ·· 1
5　Coding principles and code structures ··· 1
5.1　Coding principles ·· 1
5.2　Code structures ·· 2
6　Coding rules ··· 2
6.1　Coding rules for property owners ·· 2
6.2　Classification and coding rules for inventory ··· 3
6.3　Rules for assigning sequence codes ·· 20
Annex A (informative)　**An example of inventory codes of land requisition for hydropower projects** ·· 21

Foreword

This standard is drafted in accordance with the rules given in the GB/T 1.1-2009 *Directives for standardization—Part 1: Structure and drafting of standards*.

National Energy Administration of the People's Republic of China is in charge of the administration of this standard. China Renewable Energy Engineering Institute has proposed this standard and is responsible for its routine management. Energy Sector Standardization Technical Committee on Hydropower Planning, Resettlement and Environmental Protection is responsible for the explanation of specific technical contents. Comments and suggestions in the implementation of this standard should be addressed to:

    China Renewable Energy Engineering Institute
    No. 2 Beixiaojie, Liupukang, Xicheng District, Beijing 100120, China

Drafting organizations:

    China Three Gorges Corporation

    China Renewable Energy Engineering Institute

    POWERCHINA Beijing Engineering Corporation Limited

    POWERCHINA Kunming Engineering Corporation Limited

    POWERCHINA Huadong Engineering Corporation Limited

    Network and Information Center of Changjiang Water Resources Commission

Chief drafting staff:

JIN Heping	YAO Yingping	JI Peihuan	MENG Fanfan
ZHOU Jingliang	WEN Liangyou	YAO Yuanjun	ZHANG Jun
ZHANG Guoping	LONG Yang	LIAO Guihua	XU Pan
XIAO Yinsong	YIN Xianjun	JIANG Zhengliang	ZHANG Lin
ZHENG Yong	ZHANG Yanping	LUO Yi	LIU Can
LI Huaquan	ZHU Qiang	ZHANG Yang	LIU Yuying

Introduction

To meet the needs of digital informatization for resettlement of hydropower projects, standardize the inventory classification and coding of land requisition for hydropower projects, meet demands for sharing and utilization of the inventory of land requisition for hydropower projects, and better serve the land requisition and resettlement of hydropower projects, this standard has been developed according to the requirements of Document GNKJ [2015] No. 283, "Notice on Releasing the Development and Revision Plan of the Energy Sector Standards in 2015", issued by National Energy Administration of the People's Republic of China.

Specification for classification and coding of inventory of land requisition for hydropower projects

1 Scope

This standard specifies the basic provisions, coding principles, code structure, and coding rules of inventory of land requisition for hydropower projects.

This standard is applicable to the coding of inventory information of land requisition for hydropower projects.

2 Normative references

The following referenced document are indispensable for the application of this document. For dated references, only the dated version applies to this document. For undated references, the latest version (including any amendments) applies to this document.

GB/T 2260, *Codes for administrative divisions of the People's Republic of China*

GB/T 10114, *Rules of code representation of administrative divisions under countries*

NB/T 10102, *Code for inventory survey and census of land requisition for hydropower projects*

3 General provisions

3.1 This standard is developed to unify the requirements for classification and coding of inventory of land requisition for hydropower projects, to improve the digital informatization level of land requisition and resettlement for hydropower projects.

3.2 The inventory classification shall comply with NB/T 10102.

4 Basic requirements

4.1 The inventory code is a set of code elements, assigned to the inventory of land requisition for hydropower projects with the symbols that have certain rules and are easy to be identified and processed by computer and human. The inventory code shall contain the information on property owner, inventory classification, and sequence.

4.2 The inventory codes shall be assigned by class and grade considering the characteristics of specific inventory and the requirements of land requisition and resettlement.

4.3 The inventory codes to be increased shall be included as extended items in a unified way according to the actual situation.

5 Coding principles and code structures

5.1 Coding principles

5.1.1 An inventory code shall be unique, applicable, normative, and concise.

5.1.2 Each object of the inventory shall correspond to one code only, and one code shall identify one object only. Once an inventory code is generated, it shall be in force permanently and shall not be changed.

5.1.3 Appropriate redundancy shall be reserved for further code extension. To extend an undefined code before extended item, the code segment shall be added in a unified sequence at the end of the code that has been determined.

5.2 Code structures

5.2.1 Code structure of inventory

The code structure of inventory (Figure 1) shall consist of 3 segments. Segment 1 represents the property owner, which is a 20-digit code; segment 2 represents the category, which is a 3- to 8-digit code; segment 3 represents the sequence, which is a 3-digit code. These 3 segments are separated by the symbol "-". See Annex A for the example of a specific code.

If a project code needs to be added, the symbol "-" is used before the inventory code.

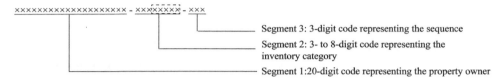

Figure 1   Code structure of inventory

5.2.2 Code structure of property owner

The code structure of property owner (Figure 2) shall consist of 2 segments. Segment 1 represents the category of property owner, which is a 2-digit code; Segment 2 represents the salient features of property owner, which is a 18-digit code.

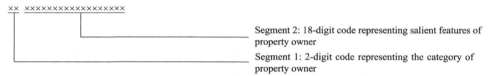

Figure 2   Code structure of property owner

5.2.3 Code structure of inventory category

The code structure of inventory category (Figure 3) shall consist of multiple layers. Layer 1 represents the primary category of inventory, which is a 2-digit code; layers 2, 3 and 4 represent the secondary category, tertiary category, and quarternary category, of inventory, respectively, which is a 1- to 6-digit code. The numbers of layers and digits of the category code are determined as per the category of the inventory, which shall not exceed 4 layers and 8 digits.

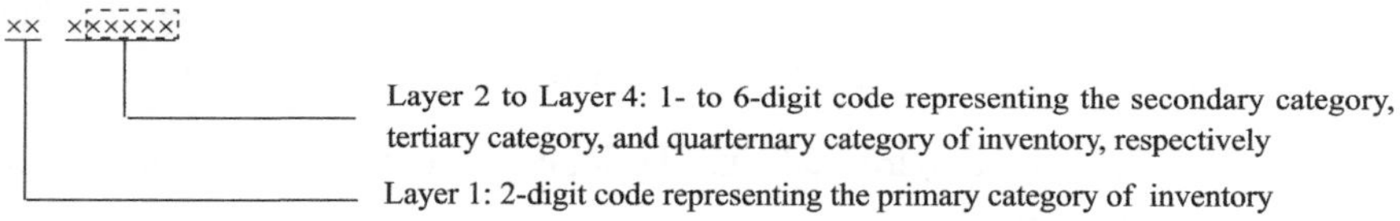

Figure 3   Code structure of inventory category

5.2.4 Code structure of sequence

The sequence is represented by a 3-digit code, ranging from 001 to 999.

6 Coding rules

6.1 Coding rules for property owners

6.1.1 Category code of property owner

The category code of property owner shall be assigned according to the three categories of residential household, collective economic organization, and administration, enterprise or public institution, respectively and shall be a 2-digit code, with 01 for residential household, 02 for collective economic organization, 03 for administration, enterprise or public institution, and 99 for ownership to be identified.

6.1.2 Salient feature code of property owner

The salient feature code of property owner (Figure 4) shall consist of 2 segments. Segment 1 represents the administration division for the location of the residential household, collective economic organization, and administration, enterprise or public institution, which is a 4-layer 15-digit code and shall comply with GB/T 2260. Layer 1 represents the administration division at the county level or above, which is a 6-digit code; layer 2 represents the street, town or township, which is a 3-digit code and shall comply with GB/T 10114; layer 3 represents the resident committee or villager committee, which is a 3-digit code ranging from 001 to 199 for resident committee or from 200 to 399 for villager committee; layer 4 represents the group or community of villager, which is a 3-digit code ranging from 001 to 999. Segment 2 represents the sequence number of the residential household, collective economic organization, and administration, enterprise or public institution within the administration division, which is a 3-digit code.

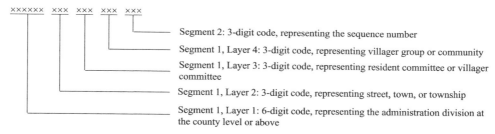

Figure 4   Salient feature code of property owner

6.2   Classification and coding rules for inventory

6.2.1   General rules for coding

Classification codes of inventory shall be assigned for the 16 categories, i.e. relocatees; houses and associated structures; land; scattered trees; rural small special facilities and agricultural and sideline facilities; individual business; cultural, educational, health, sports and religious facilities; transportation works; hydropower and water resources works; electric power works; telecommunication works; radio and television works; cultural relics and historic sites and heroes and martyrs memorial facilities; mineral resources; administrations, enterprises and public institutions; and others. The codes of these 16 categories are shown in Table 1.

6.2.2   Classification and coding for relocatees

Classification codes of relocatees shall be 3-layer 4-digit codes. Layer 1 represents the primary category of relocatees, which is 01; layer 2 represents the geographical range of the population residence, which is a 1-digit code; layer 3 represents the category of survey registration, which is a 1-digit code. Classification codes of relocatees are shown in Table 2.

Table 1   Codes for 16 categories of inventory

Code	Category	Code	Category
01	Relocatees	06	Individual business
02	Houses and associated structures	07	Cultural, educational, health, sports and religious facilities
03	Land	08	Transportation works
04	Scattered trees	09	Hydropower and water resources works
05	Rural small special facilities and agricultural and sideline facilities	10	Electric power works

Table 1 *(Continued)*

Code	Category	Code	Category
11	Telecommunication works	14	Mineral resources
12	Radio and television works	15	Administrations, enterprises and public institutions
13	Cultural relics and historic sites and heroes and martyrs memorial facilities	...	...
		99	Others

Table 2 Classification codes of relocatees

Code	Item	Code	Item
0100	Relocatees	0120	City and town relocatees
0110	Rural relocatees	0121	Residents with registered residence and housing
0111	Permanent residents with registered residence and housing	0122	Permanent residents with registered residence, but without housing
0112	Permanent residents with registered residence and housing, but temporarily transferred out	...	...
...	...	0129	Other relocatees
0119	Other relocatees		

6.2.3 Classification and coding for houses and associated structures

6.2.3.1 General rules for coding houses and associated structures

The classification codes shall be assigned for houses and their associated structures separately.

6.2.3.2 Classification and coding for houses

Classification codes of houses shall be 4-layer 7-digit codes. Layer 1 represents the primary category of houses and associated structures, which is 02; layer 2 represents the secondary category of, houses, which is 1; layer 3 represents the tertiary category of the house usage, which is a 2-digit code as shown in Table 3; layer 4 represents the structure of houses, which is a 2-digit code as shown in Table 4. Classification codes of houses are shown in Table 5.

Table 3 Codes of house usage

Code	Item	Code	Item
01	Housing	...	...
02	Outbuilding	99	Other usages

Table 4 Codes of house structures

Code	Item	Code	Item	Code	Item
01	Steel structure	04	Brick-wood structure	...	...
02	Reinforced concrete structure	05	Earth-wood structure	99	Other structures
03	Hybrid structure	06	Wood structure		

Table 5 Classification codes of houses

Code	Item	Code	Item	Code	Item
0210000	Houses	0210200	Outbuilding	0219900	Other usages
0210100	Housing	0210201	Steel structure	0219901	Steel structure
0210101	Steel structure	0210202	Reinforced concrete structure	0219902	Reinforced concrete structure
0210102	Reinforced concrete structure	0210203	Hybrid structure	0219903	Hybrid structure
0210103	Hybrid structure	0210204	Brick-wood structure	0219904	Brick-wood structure
0210104	Brick-wood structure	0210205	Earth-wood structure	0219905	Earth-wood structure
0210105	Earth-wood structure	0210206	Wood structure	0219906	Wood structure
0210106	Wood structure	...	...	...	...
...	...	0210299	Other structures	0219999	Other structures
0210199	Other structures	...	...		

6.2.3.3 Classification and coding for associated structures

Classification codes of associated structures shall be 3-layer 6-digit codes. Layer 1 represents the primary category of the houses and associated structures, which is 02; layer 2 represents the secondary category of, associated structures, which is 2; layer 3 represents the tertiary category of associated structures, which is a 3-digit code. Classification codes of associated structures are shown in Table 6.

Table 6 Classification codes of associated structures

Code	Item	Code	Item	Code	Item
022000	Associated structures	022005	Threshing ground	022010	Biogas digester
022001	Stove	022006	Water pool	022011	Well
022002	Gatehouse	022007	Vault	022012	Terrace
022003	Wall	022008	Water pipe	...	...
022004	Floor	022009	Cesspit	022999	Other associated structures

6.2.4 Classification and coding for land

Classification codes of lands shall be 4-layer 6-digit codes. Layer 1 represents the primary category of land, which is 03; layer 2 represents the secondary category of land usage, which is a 1-digit code; layer 3 represents the tertiary category of land use status, which is a 1-digit code; layer 4 represents the quarternary category of land use status, which is a 2-digit code. Classification codes of lands are shown in Table 7.

Table 7 Classification codes of lands

Code	Item	Code	Item
030000	Land	031101	Paddy field
031000	Agricultural land	031102	Irrigated land
031100	Cultivated land	031103	Dry land

Table 7 *(Continued)*

Code	Item	Code	Item
031200	Garden land	032105	Land for commercial and financial purposes
031201	Orchard	032106	Land for recreation purpose
031202	Tea garden	032199	Land for other commercial services
031203	Rubber plantation	032200	Land for industry, mining and warehousing
031299	Other garden land	032201	Industrial land
031300	Forest land	032202	Land for mining
031301	Arbor forest land	032203	Land for salt pan
031302	Bamboo forest land	032204	Warehouse land
031303	Mangrove forest land	032300	Land for residential buildings
031304	Forest swamp	032301	Urban residential land
031305	Shrub land	032302	Rural residential land
031306	Swamp in shrub forest	032400	Land for public management and public service
031399	Other forest land	032401	Land for governmental institutions and organizations
031400	Grassland	032402	Land for press and publication house
031401	Natural grassland	032403	Land for educational purpose
031402	Swamp grassland	032404	Land for scientific research purpose
031403	Artificial grassland	032405	Medical and health land
031500	Land for transportation	032406	Land for social welfare
031501	Rural roads	032407	Land for cultural facilities
031600	Land for water area and water resources facilities	032408	Sport use land
031601	Reservoir surface	032409	Land for public facilities
031602	Surface of Water pools	032410	Land for park and greening space
031603	Ditch	032500	Land used for special purpose
031900	Other land	032501	Land for military facilities
031901	Agricultural facility land	032502	Land for embassies and consulates
031902	Ridge of field	032503	Land for penitentiaries
032000	Construction land	032504	Land for religious activities
032100	Land for commercial services	032505	Funeral land
032101	Land for retail business	032506	Land for scenic facilities
032102	Land for wholesale market	032600	Land for transportation
032103	Land for catering purposes	032601	Land for railway lines
032104	Land for hotels	032602	Urban rail transit land

Table 7 *(Continued)*

Code	Item	Code	Item
032603	Land for highway	033200	Land for water area and water conservancy facilities
032604	Land for roads in towns and villages	033201	River surface
032605	Land for transportation service facilities and stations	033202	Lake surface
032606	Land for airports	033203	Coastal beach
032607	Land for harbors and wharves	033204	Inland beach
032608	Land for pipeline transportations	033205	Swamp
032700	Land for water area and water conservancy facilities	033206	Glaciers and firn
032701	Land for hydraulic construction	033900	Other land
032900	Other land	033901	Saline and alkaline land
032901	Idle land	033902	Sandy land
033000	Unused land	033903	Bare land
033100	Grassland	033904	Bare rocky and gravelly land
033199	Other grasslands		

6.2.5 Classification and coding for scattered trees

Classification codes of scattered trees shall be 4-layer 8-digit codes. Layer 1 represents the primary category of the scattered trees, which is 04; layer 2 and layer 3 represent the usage and species of the scattered trees, respectively, which are 1-digit code and 3-digit code as shown in Table 8; layer 4 represents the sizes of the trees, which is a 2-digit code as shown in Table 9. Classification codes of scattered trees are shown in Table 10.

Table 8 Codes of scattered trees by usage and species

Code	Item	Code	Item	Code	Item
1000	Economic tree	2005	Cinnamomum camphora	3007	Pomelo
1001	Vernicia fordii	2006	Ailanthus altissima	3008	Citron
1002	Oil-seed camellia	2007	Melia azedarach	3009	Jelly peach
1003	Shellac	...	...	3010	Walnut
1004	Medicinal plants	2999	Other timber trees	3011	Chinese chestnut
...	...	3000	Fruit tree	...	...
1999	Other economic trees	3001	Musa basjoo	3999	Other fruit trees
2000	Timber tree	3002	Banana	4000	Landscape tree
2001	Bamboo	3003	Peach	4001	Ginkgo
2002	Fir	3004	Plum	...	...
2003	Pine	3005	Pear	4999	Other landscape trees
2004	Cupressus funebris	3006	Tangerine		

Table 9 Codes of scattered trees by sizes

Code	Item	Code	Item
01	Mature tree	...	...
02	Sapling	99	Others

Table 10 Classification codes of scattered trees

Code	Item	Code	Item	Code	Item
04000000	Scattered trees	04200000	Timber tree	04200602	Sapling
04100000	Economic tree	04200100	Bamboo	...	...
04100100	Vernicia fordii	04200101	Mature tree	04200699	Others
04100101	Mature tree	04200102	Sapling	04200700	Melia azedarach
04100102	Sapling	...	...	04200701	Mature tree
...	...	04200199	Others	04200702	Sapling
04100199	Others	04200200	Fir	...	...
04100200	Oil-seed camellia	04200201	Mature tree	04200799	Others
04100201	Mature tree	04200202	Sapling	...	...
04100202	Sapling	...	...	04299900	Other timber trees
...	...	04200299	Others	04299901	Mature tree
04100299	Others	04200300	Pine	04299902	Sapling
04100300	Shellac	04200301	Mature tree	...	...
04100301	Mature tree	04200302	Sapling	04299999	Others
04100302	Sapling	...	...	04300000	Fruit tree
...	...	04200399	Others	04300100	Musa basjoo
04100399	Others	04200400	Cupressus funebris	04300101	Mature tree
04100400	Medicinal plants	04200401	Mature tree	04300102	Sapling
04100401	Mature tree	04200402	Sapling	...	...
04100402	Sapling	...	...	04300199	Others
...	...	04200499	Others	04300200	Banana
04100499	Others	04200500	Cinnamomum camphora	04300201	Mature tree
...	...	04200501	Mature tree	04300202	Sapling
04199900	Other economic trees	04200502	Sapling	...	...
04199901	Mature tree	...	...	04300299	Others
04199902	Sapling	04200599	Others	04300300	Peach
...	...	04200600	Ailanthus altissima	04300301	Mature tree
04199999	Others	04200601	Mature tree	04300302	Sapling

Table 10 (*Continued*)

Code	Item	Code	Item	Code	Item
...	...	04300800	Citron	04301202	Sapling
04300399	Others	04300801	Mature tree	...	...
04300400	Plum	04300802	Sapling	04301299	Others
04300401	Mature tree	...	...	...	...
04300402	Sapling	04300899	Others	04399900	Other fruit trees
...	...	04300900	Jelly peach	04399901	Mature tree
04300499	Others	04300901	Mature tree	04399902	Sapling
04300500	Pear	04300902	Sapling	...	...
04300501	Mature tree	...	...	04399999	Others
04300502	Sapling	04300999	Others	04400000	Landscape tree
...	...	04301000	Walnut	04400100	Ginkgo tree
04300599	Others	04301001	Mature tree	04400101	Mature tree
04300600	Tangerine	04301002	Sapling	04400102	Sapling
04300601	Mature tree	...	...	...	...
04300602	Sapling	04301099	Others	04400199	Others
...	...	04301100	Chinese chestnut	...	...
04300699	Others	04301101	Mature tree	04499900	Other landscape trees
04300700	Pomelo	04301102	Sapling	04499901	Mature tree
04300701	Mature tree	...	...	04499902	Sapling
04300702	Sapling	04301199	Others	...	...
...	...	04301200	Jujube	04499999	Others
04300799	Others	04301201	Mature tree		

6.2.6 Classification and coding for rural small special facilities and agricultural and sideline facilities

6.2.6.1 General rules for coding rural small special facilities and agricultural and sideline facilities

The classification codes of rural small special facilities and agricultural and sideline facilities shall be assigned respectively.

6.2.6.2 Classification and coding for rural small special facilities

Classification codes of rural small special facilities shall be 4-layer 6-digit codes. Layer 1 represents the primary category of rural small special facilities and agricultural and sideline facilities, which is 05; layer 2 represents the secondary category of rural small special facilities, which is 1; layer 3 represents the tertiary category of rural small special facilities, which is a 1-digit code and divided into five categories: farmland irrigation and water resources facilities, water supply facilities, hydropower stations, power distribution facilities and transportation facilities;

layer 4 represents the quarternary category of rural small special facilities, which is a 2-digit code. Classification codes of rural small special facilities are shown in Table 11.

Table 11 Classification codes of rural small special facilities

Code	Item	Code	Item
051000	Rural small special facilities	051299	Other water supply facilities
051100	Water resources facilities for farmland	051300	Hydropower station
051101	Reservoir	051400	Power transmission and distribution facilities
051102	Pumping station	051401	Low voltage line
051103	Flood control (river protection or drainage) dike	051402	Distribution cabinet (board), electricity meter, etc.
051104	Aqueduct	...	...
051105	Channel	051499	Other power transmission and distribution facilities
051106	Water cellar	051500	Transportation facilities
...	...	051501	Farm machinery road
051199	Other water resources facilities for farmland	051502	Footpath
051200	Water supply facilities	051503	Bridge
051201	Pumping station	...	...
051202	Water pipe	051599	Other transportation facilities
051203	Water pool	...	...
051204	Well	051900	Other rural small special facilities
...	...		

6.2.6.3 Classification and coding for agricultural sideline facilities

Classification codes of agricultural and sideline facilities shall be 3-layer 6-digit codes. Layer 1 represents the primary category of rural small special facilities and agricultural and sideline facilities, which is 05; layer 2 represents the secondary category of agricultural and sideline facilities, which is 2; layer 3 represents the tertiary category of agricultural and sideline facility, which is a 3-digit code. Classification codes of agricultural and sideline facilities are shown in Table 12.

Table 12 Classification codes of agricultural and sideline facilities

Code	Item	Code	Item
052000	Agricultural and sideline facility	052005	Limekiln
052001	Waterwheel	052006	Brickkiln
052002	Watermill	...	...
052003	Water-powered trip hammer	052999	Other agricultural and sideline facilities
052004	Water-powered roller		

6.2.7 Classification and coding for individual business

Classification codes of Individual business shall be 2-layer 4-digit codes. Layer 1 represents the primary category of individual business, which is 06; layer 2 represents the scope of business, which is a 2-digit code. Classification codes of individual business are shown in Table 13.

Table 13 Classification codes of individual business

Code	Item	Code	Item
0600	Individual business	0603	Hotels and services
0601	Wholesale and retail business	...	...
0602	Catering business	0699	Other individual businesses households

6.2.8 Classification and coding for cultural, educational, health, sports and religious facilities

Classification codes of cultural, educational, health, sports and religious facilities shall be 3-layer 6-digit codes. Layer 1 represents the primary category of cultural, educational, health, sports and religious facilities, which is 07; layer 2 represents the secondary category of cultural, educational, health, sports and religious facilities, which is a 1-digit code and be divided into two categories: cultural, educational, health, sports facilities and religious facilities; layer 3 represents the tertiary category of cultural, educational, health, sports and religious facility items, which is a 3-digit code. Classification codes of cultural, educational, health, sports and religious facilities are shown in Table 14.

Table 14 Classification codes of cultural, educational, health, sports, and religious facilities

Code	Item	Code	Item
070000	Cultural, educational, health, sports and religious facilities	072000	Religious facilities
071000	Cultural, educational, health and sports facilities	072001	Temple
071001	Cultural activities room	072002	Taoist temple
071002	Kindergarten	072003	Mosque
071003	Clinic	072004	Church
071004	publicity column	072005	Ancestral temple
071005	Activity ground	072006	Oratory
071006	Fitness activity facilities	072007	Shrine
...	...	...	...
071999	Other cultural, educational, health and sports facilities	072999	Other religious facilities

6.2.9 Classification and coding for transportation works

6.2.9.1 General rules for coding transportation works

The classification codes of transportation works shall be assigned for railway works, highway works and waterway works separately.

6.2.9.2 Classification and coding for railway works

Classification codes of railway works shall be 4-layer 6-digit codes. Layer 1 represents the primary category of transportation works, which is 08; layer 2 represents the secondary category of railway works, which is 1; layer 3 represents the usage of railway works, which is a 1-digit code; layer 4 represents the grade of the railway works, which is a 2-digit code. Classification codes of railway works are shown in Table 15.

Table 15 Classification codes of railway works

Code	Item	Code	Item
081000	Railway	081303	Grade III railway line
081100	High-speed railway	081304	Grade IV railway line
081200	Inter-city railway	081400	Heavy-haul railway
081300	Mixed passenger and freight railway	...	...
081301	Grade I railway line	081900	Others
081302	Grade II railway line		

6.2.9.3 Classification and coding for highway works

Classification codes of highway works shall be 4-layer 6-digit codes. Layer 1 represents the primary category of transportation works, which is 08; layer 2 represents the secondary category of highway works, which is 2; layer 3 represents the tertiary category of highway works, which is a 1-digit code and divided into two categories: highway, bridge and culvert; layer 4 represents the grade of the facilities, which is a 2-digit code . Classification codes of highway works are shown in Table 16.

6.2.9.4 Classification and coding for waterway works

Classification codes of waterway works shall be 4-layer 6-digit codes. Layer 1 represents the primary category of transportation works, which is 08; layer 2 represents the secondary category of waterway works, which is 3; layer 3 represents the tertiary category of waterway works, which is a 1-digit code and divided into inland waterways, river ports, wharves, transfer stations etc. layer 4 represents the grade of the projects and the type of facilities of waterway works, which is a 2-digit code. Classification codes of waterway works are shown in Table 17.

Table 16 Classification codes of highway works

Code	Item	Code	Item
082000	Highway project	...	...
082100	Highway	082199	Other highway
082101	Expressway	082200	Bridge and culvert
082102	Grade I highway	082201	Grand bridge
082103	Grade II highway	082202	Big bridge
082104	Grade III highway	082203	Medium bridge
082105	Grade IV highway	082204	Small bridge
082106	Automobile service roads	082205	Culvert
082107	Farm tracks	...	...
082108	Foot pat	082299	Other bridges and culverts

Table 17 Classification codes of waterway works

Code	Item	Code	Item
083000	Waterway works	083206	Ship docks
083100	Inland waterway	...	...
083101	Class I waterway	083299	Other river port facilities
083102	Class II waterway	083300	Wharf
083103	Class III waterway	083301	Gravity wharf
083104	Class IV waterway	083302	Sheet pile wharf
083105	Class V waterway	083303	High pile wharf
083106	Class VI waterway	083304	Ramp wharf
083107	Class VII waterway	083305	Floating wharf
083200	River port	...	...
083201	Wharf	083399	Other wharf
083202	Breakwater levee	083400	Transfer station
083203	Revetment	...	...
083204	Warship platform	083900	Others
083205	Slideway		

6.2.10 Classification and coding for hydropower and water resources works

Classification codes of hydropower and water resources works shall be 3-layer 6-digit codes. Layer 1 represents hydropower and water resources works, which is 09; layer 2 represents different categories of hydropower and water resources works, which is a 2-digit code as shown in Table 18; layer 3 represents the rank of the hydropower and water resources works, which is a 2-digit code as shown in Table 19. Classification codes of hydropower and water resources works are shown in Table 20.

Table 18 Codes for different categories of hydropower and water resources works

Code	Item	Code	Item	Code	Item
01	Hydropower station	04	Sluice dam	...	...
02	Reservoir	05	Channel	99	Others
03	Pumping station	06	Flood dike		

Table 19 Codes for different ranks of hydropower and water resources works

Code	Rank	Code	Rank	Code	Rank
01	Large (1)	04	Small (1)	99	Others
02	Large (2)	05	Small (2)		
03	Medium	...	...		

Table 20 Classification codes of hydropower and water resources works

Code	Item	Code	Item	Code	Item
090000	Hydropower and water resources works	090303	Medium	090599	Others
090100	Hydropower station	090304	Small (1)	090600	Flood dike
090101	Large (1)	090305	Small (2)	090601	Large (1)
090102	Large (2)	...	...	090602	Large (2)
090103	Medium	090399	Others	090603	Medium
090104	Small (1)	090400	Sluice dam	090604	Small (1)
090105	Small (2)	090401	Large (1)	090605	Small (2)
...	...	090402	Large (2)	...	...
090199	Others	090403	Medium	090699	Others
090200	Reservoir	090404	Small (1)	...	...
090201	Large (1)	090405	Small (2)	099900	Others
090202	Large (2)	...	...	099901	Large (1)
090203	Medium	090499	Others	099902	Large (2)
090204	Small (1)	090500	Channel	099903	Medium
090205	Small (2)	090501	Large (1)	099904	Small (1)
...	...	090502	Large (2)	099905	Small (2)
090299	Others	090503	Medium	...	...
090300	Pumping station	090504	Small (1)	099999	Others
090301	Large (1)	090505	Small (2)		
090302	Large (2)	...	...		

6.2.11 Classification and coding for electric power works

Classification codes of electric power works shall be 3-layer 6-digit codes. Layer 1 represents electric power works, which is 10; layer 2 represents different categories of electric power works i.e. power transmission line and substation, which is a 2-digit code; layer 3 represents the voltage level, which is a 2-digit code. Classification codes of electric power works are shown in Table 21.

Table 21 Classification codes of electric power works

Code	Item	Code	Item
100000	Electric power work	100108	330 kV transmission line
100100	Transmission line	100109	500 kV transmission line
100101	Transmission lines below 10 kV	100110	750 kV transmission line
100102	10 kV transmission line	...	...
100103	20 kV transmission line	100199	Other transmission line
100104	35 kV transmission line	100200	Substation
100105	66 kV transmission line	100201	10 kV substation
100106	110 kV transmission line	100202	20 kV substation
100107	220 kV transmission line	100203	35 kV substation

Table 21 (Continued)

Code	Item	Code	Item
100204	66 kV substation	...	...
100205	110 kV substation	100299	Other substation
100206	220 kV substation	...	...
100207	330 kV substation	109900	Other electric power work
100208	500 kV substation		

6.2.12 Classification and coding for telecommunication works

Classification codes of telecommunication works shall be 3-layer 6-digit codes. Layer 1 represents telecommunication works, which is 11; layer 2 represents the different categories of telecommunication works, which is a 2-digit code; layer 3 represents the different subcategories of telecommunication works, which is a 2-digit code. Classification codes of telecommunication works are shown in Table 22.

Table 22 Classification codes of telecommunication works

Code	Item	Code	Item
110000	Telecommunication works	110201	Telecommunication station
110100	Lines	110202	Machine room
110101	Long-distance lines	110203	Signal tower
110102	Local lines	...	...
110103	Access lines	110299	Other facilities
...	...	...	...
110199	Other lines	119900	Other telecommunication works
110200	Facilities		

6.2.13 Classification and coding for radio and television works

Classification codes of radio and television works shall be 3-layer 6-digit codes. Layer 1 represents the primary category of radio and television works, which is 12; layer 2 represents the secondary category of radio and television works, which is a 2-digit code; layer 3 represents the tertiary category of radio and television works, which is a 2-digit code. Classification codes of radio and television works are shown in Table 23.

Table 23 Classification codes of radio and television works

Code	Item	Code	Item
120000	Radio and television works	...	...
120100	Transmitter facilities	120199	Other transmitter facilities
120101	Transmission lines	120200	Transmission facilities
120102	Towers and poles	120201	Transmission lines
120103	Receiving station	120202	Towers and poles
120104	Relay station	120203	Receiving station
120105	Ancillary equipment	120204	Relay station
120106	Buildings	120205	Ancillary equipment

Table 23 *(Continued)*

Code	Item	Code	Item
120206	Buildings	120304	Relay station
...	...	120305	Ancillary equipment
120299	Other transmission facilities	120306	Buildings
120300	Monitoring facilities	...	...
120301	Transmission lines	120399	Other monitoring facilities
120302	Towers and poles	...	...
120303	Receiving station	129900	Other radio and television works

6.2.14 Classification and coding for cultural relics and historic sites and heroes and martyrs memorial facilities

Classification codes of cultural relics and historic sites and heroes and martyrs memorial facilities shall be 3-layer 6-digit codes. Layer 1 represents the primary category of cultural relics and historic sites and heroes and martyrs memorial facilities, which is 13; layer 2 represents the secondary category of cultural relics and historic sites and heroes and martyrs memorial facilities, which is a 3-digit code as shown in Table 24; layer 3 represents the protection levels of cultural relics and historic sites and heroes and martyrs memorial facilities, which is a 1-digit code as shown in Table 25. Classification codes of cultural relics and historic sites and heroes and martyrs memorial facilities are shown in Table 26.

Table 24 Classification codes of cultural relics and historic sites and heroes and martyrs memorial facilities

Code	Item	Code	Item
100	Cultural relics and historic sites	200	Hero and martyr memorial facilities
101	Ancient cultural sites	201	Memorial cemeteries
102	Ancient tombs	202	Memorial hall
103	Ancient buildings	203	Memorial pavilions
104	Grotto temple	204	Memorial ancestral halls
105	Stone carvings	205	Memorial statues
106	Murals	206	Memorial cinerary halls
107	Important modern historical sites, representative buildings	207	Martyr's tombs
...	...	...	...
199	Other cultural relics and historic sites	299	Other hero and martyr memorial facilities

Table 25 Codes of cultural relics and historic sites and heros and martyrs memorial facilities by protection level

Code	Item	Code	Item
1	National level	4	County level
2	Provincial (autonomous region and municipality directly under the Central Government) level	...	...
3	Municipal level	9	Unclassified

Table 26 Codes of cultural relics and historic sites and heroes and martyrs memorial facilities

Code	Item	Code	Item
130000	Cultural Relics and Historic Sites and Hero and Martyr Memorial Facilities	131050	Stone carving
131000	Cultural relics and historic sites	131051	National level
131010	Ancient cultural site	131052	Provincial (autonomous region, municipality directly under the Central Government) level
131011	National level	131053	Municipal level
131012	Provincial (autonomous region, municipality directly under the Central Government) level	131054	County level
131013	Municipal level	...	...
131014	County level	131059	Unclassified
...	...	131060	Wall painting
131019	Unclassified	131061	National level
131020	Ancient tombs	131062	Provincial (autonomous region, municipality directly under the Central Government) level
131021	National level	131063	Municipal level
131022	Provincial (autonomous region, municipality directly under the Central Government) level	131064	County level
131023	Municipal level	...	...
131024	County level	131069	Unclassified
...	...	131070	Important historical sites and representative buildings in modern times
131029	Unclassified	131071	National level
131030	Ancient architecture	131072	Provincial (autonomous region, municipality directly under the Central Government) level
131031	National level	131073	Municipal level
131032	Provincial (autonomous region, municipality directly under the Central Government) level	131074	County level
131033	Municipal level	...	...
131034	County level	131079	Unclassified
...	...	...	...
131039	Unclassified	131990	Other cultural relics and historic sites
131040	Cave temple	131991	National level
131041	National level	131992	Provincial (autonomous region, municipality directly under the Central Government) level
131042	Provincial (autonomous region, municipality directly under the Central Government) level	131993	Municipal level
131043	Municipal level	131994	County level
131044	County level	...	...
...	...	131999	Unclassified
131049	Unclassified	132000	Memorial facilities for heroes and martyrs

Table 26 (Continued)

Code	Item	Code	Item
132010	Martyrs cemetery and memorial park	132051	National level
132011	National level	132052	Provincial (autonomous region, municipality directly under the Central Government) level
132012	Provincial (autonomous region, municipality directly under the Central Government) level	132053	Municipal level
132013	Municipal level	132054	County level
132014	County level	…	…
…	…	132059	Unclassified
132019	Unclassified	132060	Martyrs' Ashes Hall
132020	Memorial hall for heroes and martyrs	132061	National level
132021	National level	132062	Provincial (autonomous region, municipality directly under the Central Government) level
132022	Provincial (autonomous region, municipality directly under the Central Government) level	132063	Municipal level
132023	Municipal level	132064	County level
132024	County level	…	…
…	…	132069	Unclassified
132029	Unclassified	132070	Tombs of martyrs
132030	Monuments and memorial pavilion	132071	National level
132031	National level	132072	Provincial (autonomous region, municipality directly under the Central Government) level
132032	Provincial (autonomous region, municipality directly under the Central Government) level	132073	Municipal level
132033	Municipal level	132074	County level
132034	County level	…	…
…	…	132079	Unclassified
132039	Unclassified	…	…
132040	Memorial shrine	132990	Other memorial facilities for heroes and martyrs
132041	National level	132991	National level
132042	Provincial (autonomous region, municipality directly under the Central Government) level	132992	Provincial (autonomous region, municipality directly under the Central Government) level
132043	Municipal level	132993	Municipal level
132044	County level	132994	County level
…	…	…	…
132049	Unclassified	132999	Unclassified
132050	Commemorative statues		

6.2.15 Classification and coding for mineral resources

Classification codes of mineral resources shall be 2-layer 4-digit codes. Layer 1 represents the primary category of mineral resources, which is 14 ; layer 2 represents the secondary category of mineral resources, which is a 2-digit code. Classification codes of mineral resources are shown in Table 27.

Table 27   Classification codes of mineral resources

Code	Item	Code	Item
1400	Mineral resources	1404	Water and gas mineral resources
1401	Energy mineral resources	...	...
1402	Metal mineral resources	1499	Other mineral resources
1403	Nonmetallic minerals resources		

6.2.16 Classification and coding for administrations, enterprises and public institutions

Classification codes of administrations, enterprises and public institutions shall be 3-layer 4-digit codes. Layer 1 represents the primary category of administrations, enterprises and public institutions, which is 15; layer 2 represents the secondary category of, administrations, enterprises and public institutions, which is a 1-digit code and divided into three categories: administrative organizations, enterprises and public institutions; layer 3 represents the enterprise scale and the nature of profits made by public institutions, which is a 1-digit code. Classification codes of administrations, enterprises and public institutions are shown in Table 28.

Table 28   Classification codes of administrations, enterprises and public institutions

Code	Item	Code	Item
1500	Administration, enterprises and public institutions	1530	Enterprise
1510	Administration	1531	Large-scale
1520	Public institution	1532	Medium-scale
1521	Non-profit	1533	Small-scale
1522	For-profit	1534	Micro-scale

6.2.17 Classification and coding for others

Classification codes of others shall be 2-layer 4-digit codes. Layer 1 represents the primary category of others, which is 99; layer 2 represents the secondary category of, others, which is a 2-digit code. Classification codes of others are shown in Table 29.

Table 29   Classification codes of others

Code	Item	Code	Item
9900	Other objects	9905	Tourism facilities
9901	Military installations	9906	Oil or gas pipelines
9902	Prison	9907	Graves
9903	Survey marker	...	...
9904	Landmarks	9999	Others

6.3 Rules for assigning sequence codes

The sequence codes are assigned in sequence to represent the sequence number of inventory codes in the same category of the property owner, which are ranging from 001 to 999.

Annex A
(informative)
An example of inventory codes of land requisition for hydropower projects

The inventory code of the first house with reinforced concrete structure owned by ××× of Group 1, Gonghe Village, Jinping Town, Pingshan County, Sichuan Province:

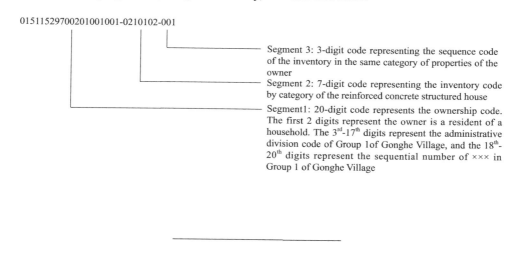